ESSENTIAL ELEMENTS

Ingredients (makes one human):   oxygen 61%, carbon 23%,
hydrogen 10%, nitrogen 2.6%, calcium 1.4%, phosphorus 1.1%,
potassium 0.2%, sulphur 0.2%, sodium 0.1%, chlorine 0.1%
plus magnesium, iron, fluorine, zinc and other trace elements.

First published in the United States of America in 2003 by
Walker Publishing Company, Inc.

Published simultaneously in Canada by Fitzhenry and Whiteside,
Markham, Ontario L3R 4T8

Printed on recycled paper

For information about permission to reproduce selections from this
book, write to Permissions, Walker & Company, 435 Hudson Street,
New York, New York 10014

Library of Congress Cataloging-in-Publication Data available
upon request

ISBN   0-8027-1408-0

Visit Walker & Company's Web site at www.walkerbooks.com

Printed in the United States of America

2   4   6   8   10   9   7   5   3   1

ESSENTIAL ELEMENTS

ATOMS, QUARKS, AND THE PERIODIC TABLE

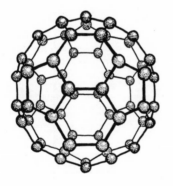

written and illustrated by

Matt Tweed

Walker & Company
New York

With love to Anna . . .

Thanks to the Blackaby family for all their support,
and as always my Mum for being great!

"*. . . that which is above is as that which is below, and that which is*
below is as that which is above, to perform the miracles of the One Thing."

The Emerald Tablet of Hermes Trismegistos

CONTENTS

Introduction	1
Early Alchemy	2
The Age of Science	4
Inside the Atom	6
Periodic Tables	8
A Burning Question	10
Bonding	12
Crystals	14
Hydrogen and Helium	16
Alkali and Alkaline-Earth Metals	18
The P-Block	20
Carbon and Silicon	22
DNA	24
Oxygen and Sulphur	26
Water and Acids	28
Halogens and the Noble Gases	30
The Transition Metals	32
The F-Block and Superheavies	34
Radioactivity	36
Orbital Structures	38
Quantum Mechanics	40
The Four Forces	42
Quarks, Leptons, and Mesons	44
Exotic Particles	46
The Big Bang	48
Stellar Fusion	50
Strings and Things	52
Constants and Hadrons	54
Carbon Chemistry	55
The Periodic Table of Elements	56
Electron Orbitals	58

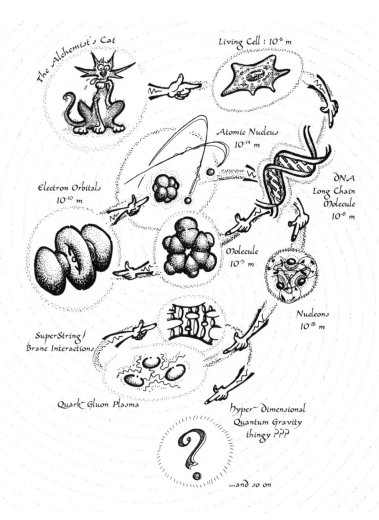

The Alchemist's Cat

Living Cell : 10^{-5} m

Atomic Nucleus
10^{-14} m

DNA
Long Chain
Molecule
10^{-8} m

Electron Orbitals
10^{-10} m

Molecule
10^{-9} m

Nucleons
10^{-15} m

SuperString /
Brane Interactions

Quark-Gluon Plasma

hyper-dimensional
Quantum Gravity
thingy ???

?

...and so on

INTRODUCTION

Pretty much all that we see or touch in our seemingly solid world is made from squintillions of tiny atoms. Combining to form the fantastic mosaic that is the visible universe, each atom is one of more than one hundred unique types of element.

If we peer closely at an individual atom, the first astonishment is that it is mainly empty space. Fizzball electrons spin complex webs around a central nucleus, a tiny point in the middle of a galaxy of whirling energy. Even here we are just scratching the surface—beyond are worlds where the rules are very strange indeed, where solidity seems to have little meaning and matter comes in waves. Zooming in, we notice that these knottings may themselves be aggregates of ever more ephemeral wisps.

The most intriguing zones live on the very edges of our understanding, dark matter that pervades the whole of space-time and the little neutrinos that zip past so fast we mainly miss them, wholesome, stable quark nuggets.

All are players in this great game of life, acts of consciousness interlacing and resonating through the all that is.

Most of all I hope you, dear reader, will enjoy this brief journey into the deep and wonderful world of matter. May we use the extraordinary knowledge we now possess with wisdom and understanding in the millenia ahead.

Brighton 2002

EARLY ALCHEMY
a wee bit of magic

The roots of chemistry stretch far back into the dim and distant past, to when our ancestors first prepared colored earths for paint, learned the secrets of fire, and started experimenting with the arcane intricacies of cookery.

The ancient Egyptians knew of seven metals, as well as carbon and sulphur, all easily extracted from natural ores. The art of *khemia,* supposedly revealed by angels, linked the metals to the seven known planets and assigned them qualities (*opposite, top left*).

Ancient Indian treatises speak of three *gunas*, fire, earth, and water. Chinese sages used two more, metal and wood (*opposite, top right*).

To the later Greek philosophers all things were made of earth, air, fire, or water (*opposite, lower left*). Naming them *elements,* Aristotle added a fifth, *quintessence*, which formed the heavens. Another philosopher, Democritus, proposed that dividing matter over and over again would eventually leave an indivisible *atmos*. Scorned by Aristotle, the *atom* was largely forgotten for centuries.

With the fall of the Greek empire, investigation of *Al-khemia* moved to Arabia. Books like Al-Razi's *The Secret of Secrets* and Jabir ibn-Hayyan's *The Sum of Perfection* told of an elixir of life that could grant immortality and transmute base metals into gold.

The quest spread to medieval Europe where alchemists like Albertus Magnus, Roger Bacon, and Nicholas Flamel hoped to find the all-powerful Gloria Mundi or Philosopher's Stone. Slowly, through experiment, trial, error, intuition, and the odd happy accident, they laid the foundations of a body of lore.

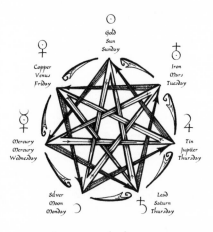

The Seven Metals of Antiquity
The Seven Known Planets

Wu-Hsing : The Five-Fold Chinese
Elemental System

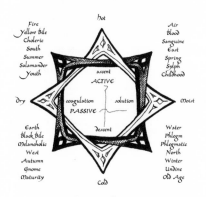

The Four Elements
and Corresponding Humors

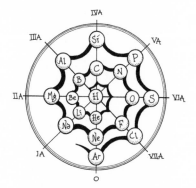

The Periodic Web of the
First Eighteen Elements

THE AGE OF SCIENCE
alchemy transmutes into chemistry

By the eighteenth century scientists were freeing themselves of metaphysical concerns, and early experiments comparing weights and masses showed many substances assumed to be elemental were in fact molecules or compounds of several parts.

In 1789 the first table of twenty-three elements was published by Antoine Lavoisier, soon followed by John Dalton's 1808 forerunner to atomic theory (which was ignored for fifty years).

As scientific techniques improved, elements were discovered at a prodigious rate. Noticing how those with similar chemical properties fell into recurring patterns, Mendeleyev created his famed *Periodic Table of Elements* in 1869, successfully predicting the existence of scandium and germanium. The first hint of stuff smaller than the atom came in 1896, when, unwittingly leaving pitchblende (a uranium ore) on an unexposed photographic plate, Becquerel accidentally discovered radioactivity.

Early twentieth-century discoveries of the surprisingly empty space around the atomic nucleus, the unveiling of the electron orbitals, and Einstein's theory that matter and energy were the same thing led Schröedinger and friends to the curiously wavy world of quantum mechanics. In 1932 the atom was split for the first time and for the rest of the century scientists explored the oddities of subatomic realms. Huge smashers hurled atoms together to synthesize new heavy elements, at other times breaking them apart to reveal whole families of exotic particles.

The universe was made of very strange things indeed.

Bronze was one of the first alloys, often cast using the lost wax process.

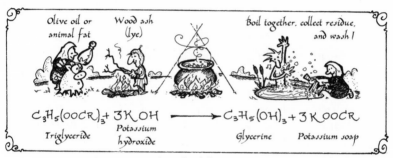

$$C_3H_5(OOCR)_3 + 3KOH \longrightarrow C_3H_5(OH)_3 + 3KOOCR$$

Triglyceride   Potassium hydroxide   Glycerine   Potassium soap

Soaps were perhaps discovered from fat falling into the dampened ashes of a fire.

Humphrey Davy
Group I and II Elements

Dimitry Mendeleyev
Periodic Table

Ramsay and Travers
Noble Gases

Marie and Pierre Curie
Radium and Polonium

Berkeley and Dubna scientists
Transuranium Elements

Some of the many involved in the discovery of the elements.

INSIDE THE ATOM
proton, neutron, and electron

A simple visualization of an atom is like a solar system (*opposite, top*), with a small central *nucleus* orbited by whirling *electrons*.

The nucleus, a mere hundred billionth of a millimeter across, contains two similar particles, *protons* and *neutrons*. Each proton has a single positive electric charge and a companion negatively charged electron. The proton count gives an element's *atomic number*, or position in the periodic table (*see pages 56–57*).

Electrons weigh nearly two thousand times less than protons and neutrons. Repelling other electrons, they are attracted to the oppositely charged protons but take no notice of neutrons (which have no charge). Balancing all these forces, electrons team up in pairs and whizz about the nucleus in *orbitals* grouped into orbital *sets*, three-dimensional patterns that get increasingly complex in larger atoms. Orbitals fill in a specific order (*opposite, below*).

Each element comes in several varieties, or *isotopes,* determined by the number of neutrons found within. Isotopes of the same element can have radically diverse chemical properties.

Incredibly, atoms are almost entirely empty space. An electron orbiting a nucleus may be visualized as a cat swinging a bumblebee on the end of a half-mile-long piece of string.

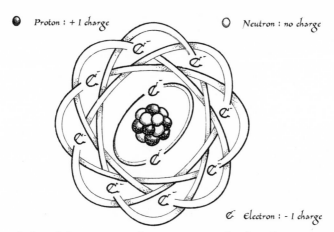

ℓ⁻ Electron : - 1 charge

The classical planetary picture of a neon atom: a central nucleus of 10 protons
and 10 neutrons surrounded by a whirl of 10 orbiting electrons.

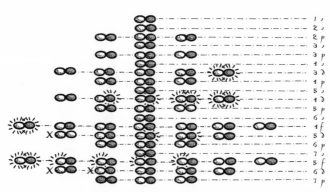

	1	s
	2	s
	2	p
	3	s
	3	p
	4	s
	3	d
	4	p
	5	s
	4	d
	5	p
	6	s
	4	f
	5	d
	6	p
	7	s
	5	f
	6	d
	7	p

Electron orbital sets build in sequence from the innermost 1 s. Each row half fills
with electrons (white blobs) before completing as oppositely spinning pairs (black blobs).
The glow around a blob indicates that one or two electrons skip to or from other orbitals,
breaking the rigid pattern. Gold, silver, and copper are among those that share this
quirk. X marks the d orbitals that try to fill before the row above gets going.

PERIODIC TABLES
elemental ordering

Each element has its own place in the periodic table. There are several versions of this, each emphasizing different features.

Professor Benfey's spiral (*below*) develops by atomic number and shows groups with the same pattern of outer electrons, and, hence, corresponding properties radiating like spokes from a hydrogen hub. As different orbitals fill, blocks form outcrops.

In contrast Dr. Stowe's table (*opposite, top*) displays the physical ordering of the intricate *shells* of orbital sets of electrons in atoms, innermost at the top, using *quantum numbers*. Panic not!

One modern version of Mendeleev's table (*opposite, bottom*) puts groups in vertical columns with horizontal *periods* of orbital sets. The elements are displayed by atomic number (the count of protons or electrons for each element), reading left to right.

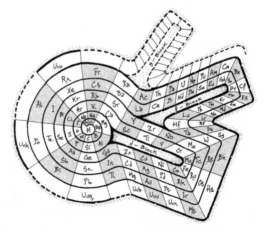

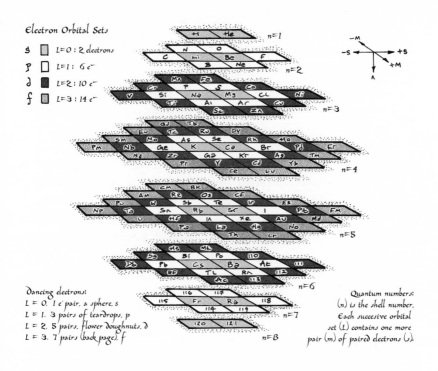

The atomic shells (above), with the shaded orbital sets of which they are comprised. Below we see the modern periodic table, which is read left to right across the open spaces and shows increasing order of atomic number (see pages 56–57).

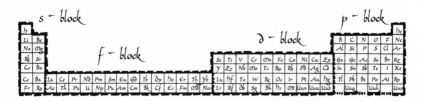

A BURNING QUESTION
chemical conflagrations

Most things around us are *compounds*, or mixtures of elements. To bond, atoms simply rearrange their outer, highest energy *valence* electrons, keeping their interiors safely out of the way.

Opposite (*top*) is an *exothermic* reaction, which produces heat. After initially loosening the gas molecules with a flame, new water bonds quickly form that lock up less energy than the original gas bonds. The released energy keeps the reaction going and the gases explode, frantically shuffling electrons.

Below we see an *endothermic* reaction that takes place when plants photosynthesize. Here the sums go the other way, and heat needs to be absorbed, in this case from sunlight. The products therefore have more energy than the reactants.

Matter itself exists in several different phases (*below*). Solids pack atoms closely in rigid arrangements. Heating makes the atoms vibrate, shaking them apart to form liquids that can flow and change shape. Heating further weakens even these loose bonds and the atoms scatter at high speed in all directions as gases. At yet higher temperatures some of the electrons are knocked off the atoms to create an electrically charged, ionized plasma, like that found in the super hot corona of the sun.

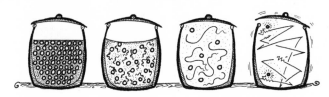

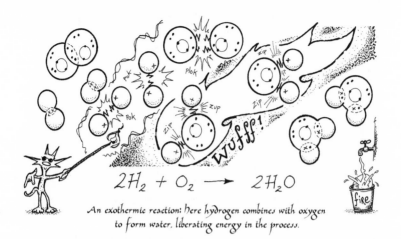

$$2H_2 + O_2 \longrightarrow 2H_2O$$

*An exothermic reaction: here hydrogen combines with oxygen
to form water, liberating energy in the process.*

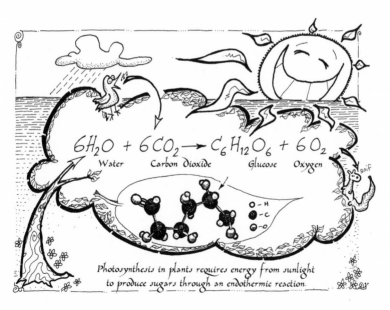

$$6H_2O + 6CO_2 \longrightarrow C_6H_{12}O_6 + 6O_2$$

Water   Carbon Dioxide   Glucose   Oxygen

○ – H
● – C
◐ – O

*Photosynthesis in plants requires energy from sunlight
to produce sugars through an endothermic reaction.*

BONDING
atomic stickiness

Molecules are formed as atoms' outer electrons share dances. Losing or gaining electrons causes atoms to become electrically charged *ions*. Most elements are either metals, which are *electropositive*, losing electrons to form *cations*; or nonmetals, which are *electronegative* and gain electrons to form *anions*.

An *ionic bond* occurs when a negative anion borrows electrons from a positive cation to give them both full orbitals like the nearest noble gas (*opposite, top left*). Though tough and brittle with high melting points, many ionic compounds dissolve in water.

Nonmetals combine using *covalent bonds,* which shuffle and share outer electrons into pairs, again filling up empty orbitals (*opposite, top right*). The attraction felt by electrons for nucleii, outweighing their mutual repulsions, holds the resulting molecule together.

In *metallic bonds* electrons float away from their nuclei, dissociating into a "sea" around a lattice of positive ions (*opposite, center*). The conductivity and shininess of metals is due to these mobile electrons, and their strength and high melting points result from the lovestruck relationship between the ions and their mates.

Hydrogen attached to a nonmetal pushes against unbonded *lone pair* electrons creating a slight charge across the molecule. If another electronegative atom is nearby, a weak *hydrogen bond*, vital in water and DNA, appears between them.

With asymmetrical motions of electrons causing instantaneous small *Van der Waals forces* between atoms, and overlapping orbitals smearing π-bonds (*bottom right*), atomic glues come in many forms.

Li⁺   Fl⁻

Lithium flouride ionically bonding.

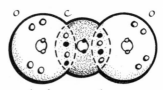

O   C   O

Covalent bonding in carbon dioxide.

Pseudo-electron density map of crystalline lithium flouride.
(nuclei centers are 200.9 pico-meters apart).

The Noble Ship of Current sails the Metallically Bonded Dissociated Sea of Electrons.

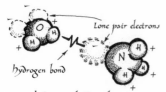

Lone pair electrons

hydrogen bond

hydrogen bonding between
water (H_2O) and
ammonia (NH_3)

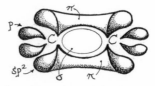

π-bonds formed from overlapping orbitals
alongside a covalent σ-bond create the
carbon double bond in ethylene.

CRYSTALS
building bigger

Crystals are simple repeated patterns of *unit cells*. Like apples in a market stall, atoms and molecules can grow into large three-dimensional solids, the pieces positioned to give the best balance between attractive and repulsive forces.

There are seven crystalline systems based on tessellating geometries that, when combined with the four basic unit cell types, give the fourteen *Bravais Lattices* (*opposite*). Variations in temperature or pressure may change one crystal structure into another that is more comfortable and efficient. Sulphur, for example, transforms from an orthorhombic lattice to a mono-clinic at 96°C, quickly reverting back on cooling.

Ordered in their zillions, these tiny blocks build into many of the solid substances of our world, bridging the vast difference of scale between molecules, minerals, and mountains.

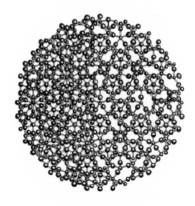

Simple Cubic

Body-Centered Cubic

Hexagonal System

Face-Centered Cubic

Simple Monoclinic

Base-Centered Monoclinic

Rhombohedral System

Body-Centered Tetragonal

Simple Tetragonal

Body-Centered Orthorhombic

Simple Orthorhombic

Face-Centered Orthorhombic

Triclinic System

Base-Centered Orthorhombic

15

HYDROGEN AND HELIUM
the first two elements

Hydrogen makes up three quarters of all known matter in the universe and is a large part of most stars. Element one, the simplest atom, consists of one proton orbited by one electron.

Hydrogen gas is *diatomic*, which means two atoms form one molecule. Highly explosive in air, it burns rapidly with oxygen to create water. Under immense pressures and temperatures—everyday conditions in the cores of giant planets like Jupiter and Saturn—hydrogen becomes metallic.

Beyond its common form, hydrogen has two isotopes, *deuterium*, which has one neutron, and *tritium*, with two. Tritium is rare and also unstable and therefore radioactive.

The second element in the periodic table has two protons, two electrons, and (99.99% of the time) two neutrons. Called *helium*, it is the second most abundant element in the material universe. With two electrons filling the *1s*-orbital, helium is happy to stay on its own and rarely reacts with other elements. It is the first of the *noble,* or *inert*, gases, each of which have completely full outer electron orbitals.

Surprisingly, helium was unknown on Earth until 1870 when it was discovered through *spectrographic analysis* of sunlight, a fingerprinting technique for elements (*opposite, bottom*).

Lighter than air, though twice as heavy as hydrogen, the specks of helium formed here quickly float off into outer space. It is a much safer gas than hydrogen in balloons, and when inhaled will produce a squeaky voice.

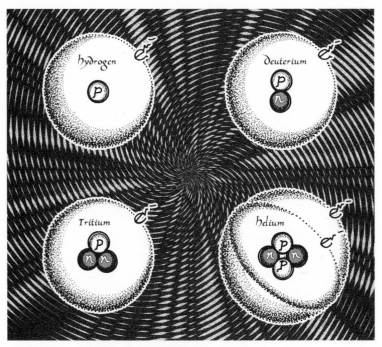

The three isotopes of hydrogen and an atom of helium
showing protons (p), neutrons (n), and electrons (e).

Each element absorbs light in a unique way, producing dark bands at
specific places across the electromagnetic spectrum: in this way even
very distant stars can be analyzed to discover their chemistry.

ALKALI AND ALKALINE-EARTH METALS
the violent world of the s-block

The first real group of the periodic table is known as the *alkali metals*, IA (*below, far-left column*). Soft, silvery white, and very electropositive, they all have a single outer *s*-orbital electron that they enthusiastically lose to form +1 ions.

Lithium, the third element, is the lightest metal and floats on water. *Sodium*, below it, floats and fizzes as it oxidizes, also commonly bonding with chlorine as salt (NaCl). Continuing down the group, *potassium* is the second lightest metal, oxidizing rapidly in air and bursting into flames when wet. *Caesium*, the most electropositive element, and rubidium explode on contact with air. *Francium*, finally, is radioactive.

Moving along the column we meet group IIA, the rare earth metals, *beryllium, magnesium, calcium, strontium, barium,* and *radium*. Slightly less electropositive, they gladly form +2 ions, losing both their outer electrons. They are denser than their group I neighbors, with higher melting and boiling points.

A wire dipped in compounds of these elements will produce characteristic colors when held in a flame. Excited electrons jump between orbitals, losing their energy as little packets of light, *photons*, on the way back down to their normal state.

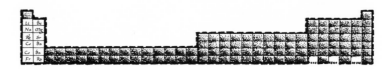

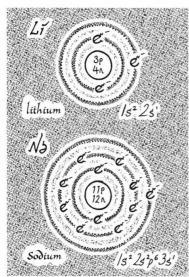

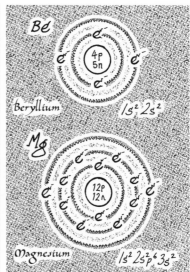

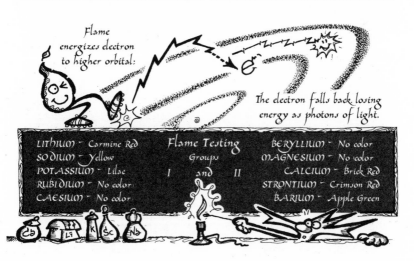

THE P-BLOCK
metals, metalloids, and nonmetals

Elements five to ten are the first members of the *p-block*. Up to six electrons inhabit three new double-teardrop shaped *p*-orbitals arranged at right angles around the nucleus (*see page 38*). They appear as solids (carbon and aluminum), liquids (bromine), and gases (nitrogen and chlorine), depending on the balance of their interatomic and intermolecular forces.

The left-hand side of the *p*-block mostly shines with solid metals. Ductile, malleable, and conductive because of their footloose outer electrons, metals can be stretched into wires, squished into sheets, or mixed together into alloys.

Crossing the block from left to right, metals transform into nonmetals. These are dull, brittle solids, liquids, or gases that are poor conductors of heat and electricity. With full or nearly full orbitals they prefer to share electrons covalently. In this small corner we find many of the major players in the game of life, such as carbon, oxygen, and nitrogen (*opposite, top*).

In between lie the metalloids, a diagonal streak of ambiguous elements with aspects of both types of behavior. Among these are the semiconductors boron, silicon, germanium, and arsenic, which form the hearts of our computers and electronic gizmos.

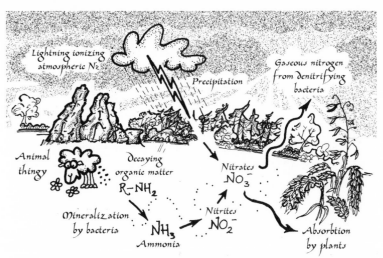

Lightning ionizing atmospheric N₂

Precipitation

Gaseous nitrogen from denitrifying bacteria

Animal thingy

Decaying organic matter
$R-NH_2$

Mineralization by bacteria
NH_3
Ammonia

Nitrites
NO_2^-

Nitrates
NO_3^-

Absorbtion by plants

To be useful to plants the strong N₂ triple bond has to be broken by nitrogen-fixing bacteria.

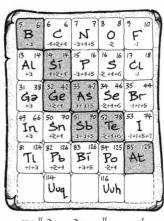

5 6	6 7	7 8	8 9	9 10
B +3	C -4+2+1	N -3+1+5	O -2	F -1
13 14 Al +3	14 14 Si -4+2+1	15 16 P -3+1+5	16 16 S -2+1+6	17 18 Cl -1
31 38 Ga +3	32 42 Ge +2+1	33 42 As +3+1+5	34 46 Se -2+1+6	35 44 Br -1+1+5
49 66 In +3	50 70 Sn +2+1	51 70 Sb +3+1+5	52 78 Te +2+1+6	53 74 I -1+1+5+7
81 124 Tl +1+3	82 126 Pb +2+1	83 126 Bi +3+5	84 125 Po -2+1	85 125 At
	114 Uuq	116 Uuh		

Metalloids run diagonally across the p-block marking the changeover between metals and nonmetals.

Boron isn't boring.
I can tell you that.
This bug isn't resting
It's quite dead out flat
All ready for the cooking in
A flameproof Pyrex dish
of
Borosilicates!!
The fifth atom's
special wish.

BORON IS NOT BORING

CARBON AND SILICON
organic and virtual thinking materials

Twenty-three percent of you is carbon. The sixth element underpins organic chemistry, the fabric of life, from DNA and proteins in our cells to once living stuff, plastics, and fossil fuels.

Coming in a dazzling array of molecules, carbon is neither electropositive nor negative. A nonmetal, it combines with many other elements and also extensively with itself, creating long chains and rings (*see page 55*). Multiple π–bonds smear electrons between atoms to give double and triple bonds.

Carbon arranges as several different *allotropes*. In diamond crystals every atom bonds to four others in a tetrahedral grid (*top right*) whereas in graphite, a soft crystalline solid found in charcoal and pencils, flat planes of carbon rings slide easily over each other (*top left*). Each atom here joins to three others, the π–bonds enabling it to conduct electricity.

Other carbon allotropes include a series of spherical *buckyball* molecules, the *Buckminsterfullerenes*, and intriguing *nanotubes* that self-assemble in the right conditions.

Directly beneath carbon in the periodic table is *silicon*, a metalloid semiconductor. Carbon life is mirrored by silicon logic in the buzzing mazes of electronic microchips, which consist of purified silicon crystals, doped with elements like gallium or arsenic to alter their conducting properties.

Stable silicon compounds cover over a quarter of the Earth as rocks and sand; and clays, silicates with unusual life-mimicking chemistries, perhaps even led to biological evolution.

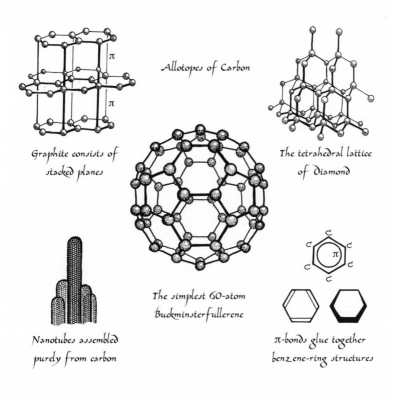

Allotopes of Carbon

Graphite consists of
stacked planes

The tetrahedral lattice
of diamond

The simplest 60-atom
Buckminsterfullerene

Nanotubes assembled
purely from carbon

π-bonds glue together
benzene-ring structures

In silicon dioxide (Si O₂) each silicon atom is tetrahedrally bonded to four
oxygen atoms to create quartz crystals and the sand on our beaches.

DNA

the elements of life

Deoxyribonucleic acid, DNA to its friends, is the incredible long-chain molecule at the heart of our cells, which stores the genetic instructions that code for the myriad bits of our bodies.

Twisted into a double helix, DNA consists of two chains of alternating sugar and phosphate molecules. Stretched between the sugars are pairs of *bases (opposite, bottom)* joined by hydrogen bonds *(below)*. Being fussy, adenine (A) only bonds to thymine (T), and cytosine (C) only to guanine (G), making the two strands exact complementary opposites of each other.

When our cells divide, a precise copy has to be made of our twenty-three pairs of chromosomes, each a DNA molecule of some 250 million base pairs. Replication begins with the helix unwinding, which allows two new strands to form on the exposed bases. These then curl into two separate molecules, each with one old strand and one new *(main picture opposite)*.

To read genetic plans, the DNA undoes a few turns, allowing *messenger RNA,* ribonucleic acid, a close relative of DNA that uses uracil (U) instead of thymine, to copy bases by hydrogen bonding. Itself uncoiling, the mRNA then carries the code to molecular factories elsewhere in the cell to make the required part, whether bone or paw, petal or claw.

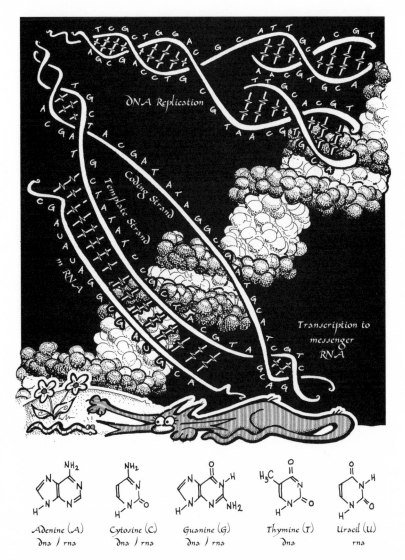

DNA Replication

Coding Strand

Template Strand

mRNA

Transcription to messenger RNA

Adenine (A)
dna / rna

Cytosine (C)
dna / rna

Guanine (G)
dna / rna

Thymine (T)
dna

Uracil (U)
rna

OXYGEN AND SULPHUR
the over and underworlds of group VI A

A fifth of the air we breathe is *oxygen*. After hydrogen and helium, it is the third most abundant element in the universe. A highly reactive gas, needing only two more electrons to fill its outer orbitals, it plays a considerable role on our planet.

Usually diatomic (O_2), the eighth element also comes in threes as *ozone* (O_3). Found naturally in the upper atmosphere protecting us from cosmic radiation, ozone's tendency to *oxidize*, or add oxgen atoms to, many ions means that nasty chemicals high in the air can leech away this precious shield.

Just under half the Earth's crust consists of oxygen atoms. The ten most common compounds are all *oxides*; just under half is sand, silicon dioxide (SiO_2), a third is magnesium oxide (MgO), and much of the rest is iron(II) oxide (FeO). Water (H_2O), that most useful stuff, is another oxygen compound.

Below oxygen in the periodic table is the underworld of *sulphur*. Usually found as a brittle, pale yellow solid, it displays a profusion of multi-atom ring-and-chain allotropes, burning in air to create sulphur dioxide (SO_2). Combining with water in clouds, it can become tree-destroying sulphurous acid rain.

Sulphur is less electronegative than oxygen, and hydrogen sulphide (H_2S) acts differently than water, hydrogen bonding having little influence. To us a fiendishly toxic gas with a rotten-egg smell, whole colonies of creatures nevertheless live in the dark on energy metabolized by sulphur-breathing bacteria beside deep ocean volcanic vents bubbling hot H_2S.

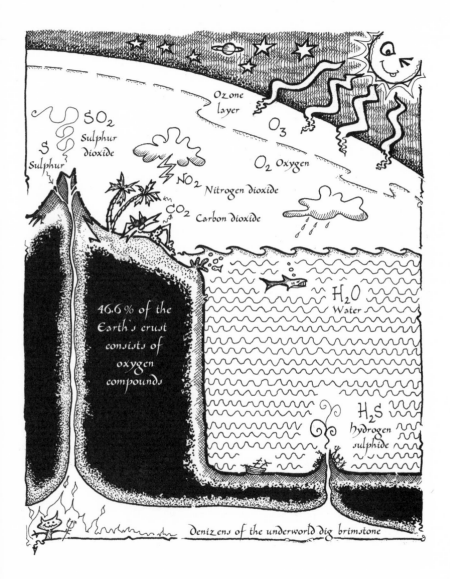

Ozone layer

O_3

SO_2
Sulphur dioxide

S
Sulphur

O_2 Oxygen

NO_2 Nitrogen dioxide

CO_2 Carbon dioxide

46.6 % of the Earth's crust consists of oxygen compounds

H_2O
Water

H_2S
hydrogen sulphide

Denizens of the underworld dig brimstone

WATER AND ACIDS
making a splash

Water is the most common molecule in the universe. One oxygen and two hydrogen atoms, H_2O is two-thirds of you.

The water molecule is *polarized*. The pull from the oxygen atom gives the hydrogen atoms a slight positive charge (*opposite, top left*) and extensive networks of hydrogen-bonded molecules carve sixfold snow crystals (*opposite, center*), cause surface tension, and create the fluctuating liquid crystal lattice we drink (*below*).

Water's polarity allows it to easily unsettle and dissolve other polarized molecules into ions, water itself dissociating into H^+ protons and *hydroxides*, OH^-. *Acids* are compounds that donate protons in solution, attacking metals to liberate hydrogen gas. *Bases* are compounds that accept protons, and include metal oxides, hydroxides, and amines. Soapy and bitter, they combine with an acid to form a salt and water. For example, hydrochloric acid and sodium hydroxide make salt and water: $HCl + NaOH = NaCl + H_2O$. Other *Lewis* acids and bases respectively accept or donate an electron pair subject to the solvents used, and not all acids need water.

Two slightly different acid reactions power the electron flows of the batteries (*opposite, bottom*) that light up Mr. Frosty.

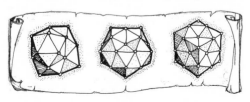

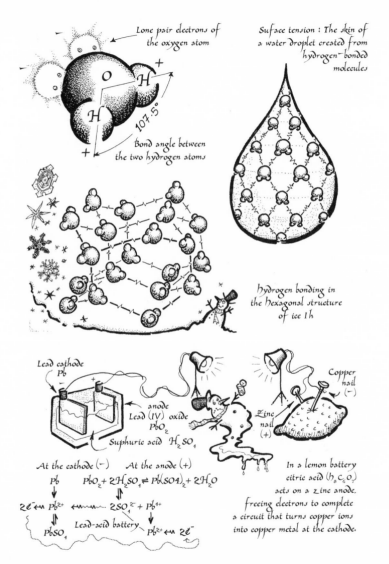

Lone pair electrons of the oxygen atom

O
H +
H
107.5°
+

Bond angle between the two hydrogen atoms

Suface tension: The skin of a water droplet created from hydrogen-bonded molecules

Hydrogen bonding in the hexagonal structure of ice Ih

Lead cathode Pb

anode Lead (IV) oxide PbO_2

Suphuric acid H_2SO_4

Copper nail (–)

Zinc nail (+)

At the cathode (–)   At the anode (+)

Pb $PbO_2 + 2H_2SO_4 \rightleftharpoons Pb(SO4)_2 + 2H_2O$

$2e^- \leftarrow Pb^{2+}$ $2SO_4^{2-} + Pb^{4+}$

$PbSO_4$ Lead-acid battery $Pb^{2+} \leftarrow 2e^-$

In a lemon battery citric acid $(h_8c_6o_7)$ acts on a zinc anode, freeing electrons to complete a circuit that turns copper ions into copper metal at the cathode.

HALOGENS AND THE NOBLE GASES
ups and downs at period's end

The universe's most inert and vigorous elements are found in the final two columns of the periodic table. The members of group VIIA, the *halogens*, are just one electron short of a full shell, and aggresively form compounds. All elements, except helium, neon, and argon, bond with a halogen to form a *halide*.

The ninth element, and easily the most electronegative, is *fluorine*, a pale green-yellow diatomic gas that combines fanatically with almost anything, attacking compounds to form *fluorides*. The rest of the group are also intensely reactive, particularly that rascal chlorine (*see mountain peaks opposite*).

With one more proton and one more electron added, we finally meet the quiet, solipsistic group VIIIA. With all the places of their electron orbitals full, they are closed to business and, on the whole, content not to react with anything.

That said, *xenon* does form (with effort) a few compounds with feisty fluorine and its neighbor, oxygen, and a few helium and krypton compounds also exist, so the former name of this group, the *inert gases*, was changed to the slightly less lazy *noble gases*. When you next see glowing neon signs, picture those full orbitals frantically buzzing with jumping electrons.

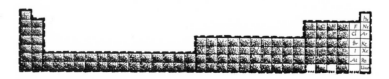

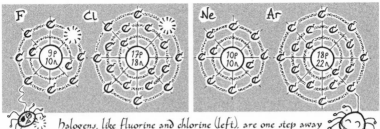

halogens, like fluorine and chlorine (left), are one step away
from the stable, full electron orbitals of their neighbors,
the noble gases neon and argon (right).

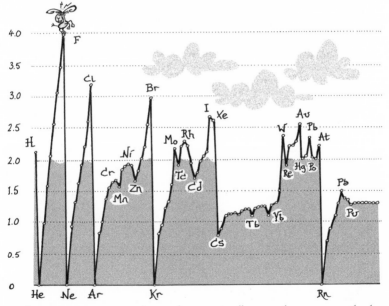

Electronegativity measures how easily an atom will attract electrons in a molecule:
the very reactive halogens occupy the highest peaks while the noble gases
(with the notable exception of xenon) sleep quietly in the deepest valleys.

THE TRANSITION METALS
gold, silver, copper, and iron in the d-block

The next zone of the periodic table we encounter is a series of metals starting at scandium where the first of ten electrons begins filling the set of *3d* orbitals inside the *4s*. Most members of this series heartily lose one or more electrons to form a bewildering array of brightly colored compounds.

The *transition metals* are hard and strong, their similar structures allowing them to be mixed into useful *alloys*. Copper and zinc combine into brass, and mercury, the only metal liquid at room temperature, forms alloys called *amalgams*.

With a dash of carbon, iron creates steel, becoming even harder with an added splash of vanadium, molybdenum, or chromium. Iron's attraction is due to the unbalanced magnetic moments of unpaired electrons in its outer *d*-orbital. Several neighbors also exhibit varying degrees of paramagnetism.

Titanium has a reputation for both strength and corrosive resistance, and is thus ideal for flying machines and rocket ships.

The enduring popularity of shiny gold, silver, and copper is in part due to their stability. Excellent conductors of electrons and heat, they have many applications in electronics and optics, also looking pretty in rings, crowns, coins, and other baubles.

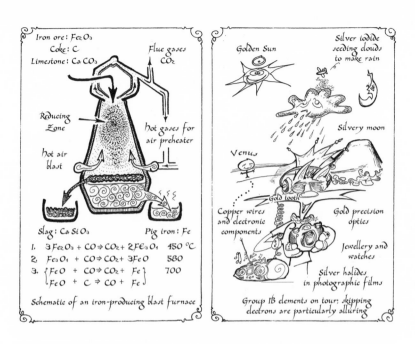

The text in the left diagram:
Iron ore: Fe_2O_3
Coke: C
Limestone: $CaCO_3$
Flue gases CO_2
Reducing Zone
hot gases for air preheater
hot air blast
Slag: $CaSiO_3$
Pig iron: Fe

1. $3Fe_2O_3 + CO \Rightarrow CO_2 + 2Fe_3O_4$ 450 °C
2. $Fe_3O_4 + CO \Rightarrow CO_2 + 3FeO$ 580
3. $\begin{cases} FeO + CO \Rightarrow CO_2 + Fe \\ FeO + C \Rightarrow CO + Fe \end{cases}$ 700

Schematic of an iron-producing blast furnace

The text in the right drawing:
Golden Sun
Silver iodide seeding clouds to make rain
Silvery moon
Venus
Copper wires and electronic components
Gold tooth
Gold precision optics
Jewellery and watches
Silver halides in photographic films

Group IB elements on tour: skipping electrons are particularly alluring

21 1-2 **Sc** +3	22 2-2 **Ti** +2+3+4	23 3-2 **V** +2+3+4+5	24 5-1 **Cr** +2+3+6	25 5-2 **Mn** +2+3+4+7	26 6-2 **Fe** +2+3	27 7-2 **Co** +2+3	28 8-2 **Ni** +2+3	29 10-1 **Cu** +1+2	30 10-2 **Zn** +2
39 1-2 **Y** +3	40 2-2 **Zr** +4	41 4-1 **Nb** +3+5	42 5-1 **Mo** +6	43 5-2 **Tc** +4+6+7	44 7-1 **Ru** +3	45 8-1 **Rh** +3	46 10-0 **Pd** +2+4	47 10-1 **Ag** +1	48 10-2 **Cd** +2
57 1-2 **La** +3	72 2-2 **Hf** +4	73 3-2 **Ta** +5	74 4-2 **W** +6	75 5-2 **Re** +4+6+7	76 6-2 **Os** +3+4	77 7-2 **Ir** +3+4	78 9-1 **Pt** +2+4	79 10-1 **Au** +1+3	80 10-2 **Hg** +1+2
89 1-2 **Ac** +3	104 ? **Rf** +4	105 ? **Db** ?	106 ? **Sg** ?	107 ? **Bh** ?	108 ? **Hs** ?	109 ? **Mt** ?	110 ? **Uun** ?	111 ? **Uuu** ?	112 ? **Uub** ?

The d block transition metals, showing for each element the number of protons (top left), the number of electrons in the outer d- and s-orbitals (top right), and oxidation states (below). Electrons skip from outer s to d-orbitals when the latter are half-full or full.

THE F-BLOCK AND SUPERHEAVIES
enormous atoms and islands of stability

At lanthanum, element fifty-seven, a *5d* orbital starts to fill before something strange happens: the next electron drops into a previously hidden *4f* orbital inside the full *6s, 5s,* and *5p* orbital sets, taking the electron from the *5d* with it. The *5d* orbitals wait patiently until the *4f*s are full, apart from one halfway hiccup at gadolinium, where an electron flickers up to the *5d.*

Quietly spreading from lanthanum to lutetium the *lanthanides,* or *rare-earth* metals, fill a fourteen place set of *4f* orbitals. Overshadowed by their *5s* and *5p* sets, only subtle chemical differences are found in this series.

The radioactive *actinides* play much the same trick, as two electrons begin a *6d* orbital only to quit the job and turn within to fill a *5f* orbital instead. Uranium is the last natural element; artificially made atoms now fill a seventh of the periodic table.

Beyond the *f*-block at rutherfordium, a fourth transition series starts filling a *6d* orbital. These superheavy elements tend to be highly radioactive and unstable. Elements with up to 118 protons have been fleetingly created in particle accelerators.

Around elements 114 "eka-lead" and 184 there may possibly be rare islands of stability where a few isotopes with balanced nucleii have significant lifetimes of minutes rather than seconds.

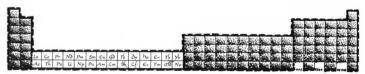

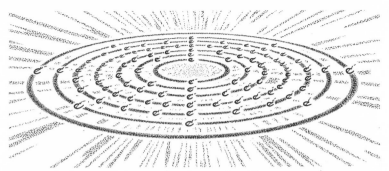

The electrons of plutonium (not the order they fill in: see page 7):
$1s^2$ $2s^2$ $2p^6$ $3s^2$ $3p^6$ $3d^{10}$ $4s^2$ $4p^6$ $4d^{10}$ $4f^{14}$ $5s^2$ $5p^6$ $5d^{10}$ $5f^6$ $6s^2$ $6p^6$ $7s^2$

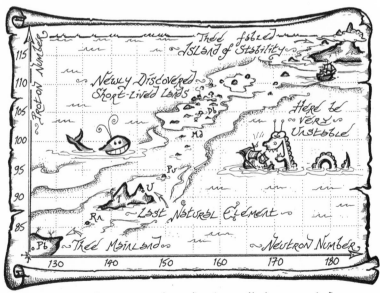

A Mappe to show ye wayye to lands afar whence stable elementes might forme wythe a balancing of neutron and protonne numbers

35

RADIOACTIVITY
nuclear fizzicks

An atomic nucleus, held together by immensely strong forces, contains huge amounts of energy. Unstable nuclei rebalance by spitting out radioactive emissions as protons and neutrons join (*fusion*) or split off (*fission*). An isotope's radioactivity halves over its *half-life*, the less stable it is the faster it *decays*. Uranium–238 has a half-life of 4.5 billion years, yet with ten neutrons less, uranium–228 lasts a mere fifth of a second!

All living things are slightly radioactive, constantly absorbing carbon–14 and tritium generated by cosmic rays. At death, we stop gathering these isotopes, and archaeologists use the 5730 year half-life of carbon–14 to date interesting goo.

Beyond bismuth, all elements have *radioisotopes* that undergo α-*(alpha)* decay, the nucleus expelling an α-*particle* (a helium nucleus, *see page 17*). Thin clothing should prevent α-particles from ionizing you. Excess neutrons in a nucleus cause β-*(beta)* decay, a neutron converts into a proton, releasing a speeding electron (a β-*particle*); protective apparel or 2 mm of aluminum halts these beasties. Often found alongside α or β-decays, γ-*(gamma) rays* carry off energy as electromagnetic radiation that can zip through up to 8 inches of lead with no difficulty.

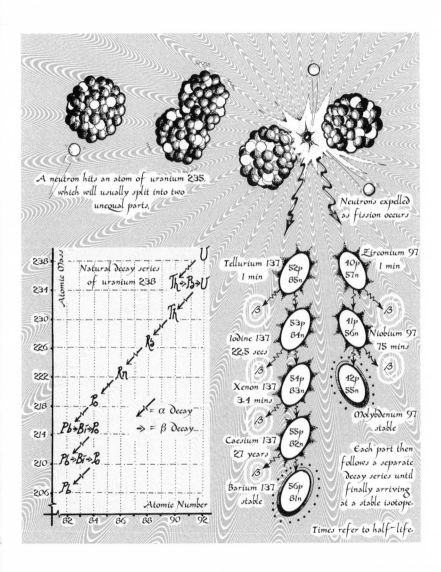

A neutron hits an atom of uranium 235, which will usually split into two unequal parts,

Neutrons expelled as fission occurs

Natural decay series of uranium 238

Th→B→U

Atomic Mass

U

Th

Ra

Rn

Po

Pb→Bi→Po

Pb→Bi→Po

Pb

↖ = α decay

→ = β decay

Atomic Number

82   84   86   88   90   92

Tellurium 137
1 min

52p
85n

β

Iodine 137
22.5 secs

53p
84n

β

Xenon 137
3.4 mins

54p
83n

β

Caesium 137
27 years

55p
82n

β

Barium 137
stable

56p
81n

Zirconium 97
1 min

40p
57n

β

Niobium 97
75 mins

41p
56n

β

Molybdenum 97
stable

42p
55n

Each part then follows a separate decay series until finally arriving at a stable isotope.

Times refer to half-life.

37

ORBITAL STRUCTURES
the whirly world of the very small

At the scale of fundamental particles like the electron, energy comes in little bits, or *quanta*, and things down here behave like both particles and/or waves, depending on your perspective.

If an atom is pumped up with energy, electrons whizzing around the nucleus excitedly make discrete *quantum* jumps into new orbitals that fit together like a buzzing ethereal flower.

Mathematical *wave functions* can predict the probability of finding an electron in a specific place. Nine times out of ten the electron will be found within the density plot (*opposite, left*). Some of the time it's possibly on the other side of the galaxy.

Each orbital may be inhabited by two electrons, with opposite spins to each other. The sphere (*opposite, right*) represents a *1s* orbital and the twins may be anywhere, including in the nucleus. The next orbital set, the *2p*, fills as shown below. Three double tear-drop shapes reflect around a nuclear *nodal plane* where electrons hardly ever go. As further orbitals fill, new electrons are forced into more and more exotic dances (*opposite, bottom, and see page 58*).

Strangely, the wavy nature of electrons means that we can't know their position *and* speed at the same time. A small but measurable uncertainty creeps in. Just looking for something so tiny radically alters its behavior.

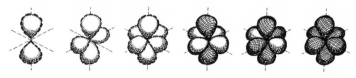

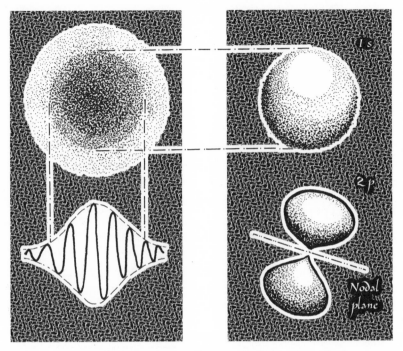

Finding the elusive electron : from wave packet to atomic orbital.

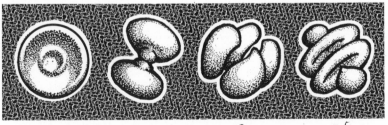

2s   3p   4d   5f

An assortment of orbital shapes : note that only s orbitals lack a nodal plane.

QUANTUM MECHANICS
curiouser and curiouser

Deep inside the *nucleons,* the protons and neutrons, lurks an even smaller realm. Here energy and matter have such a close relationship that it's sometimes difficult to tell them apart.

Each nucleon is made of three *quarks*, particlelike matter fields that are the building blocks of the everyday universe. The quarks we usually meet are called *up* and *down*.

Peeking in a proton (*top left*) we find two up quarks and one down quark. Up quarks have an electric charge of plus two-thirds, while downs carry minus one-third. Adding the three quark charges gives the proton's total of plus one. The neutron has a slightly different crew of two down quarks and an up, which cancel out, leaving it electrically neutral (*top right*).

Binding quarks together is the job of the *nuclear strong force.* Instead of the two charges of electricity, the strong force has three charges to balance, which we can liken to the three primary colors of light. If we mix red, green, and blue lights we get a neutral white. Similarly, each nucleon carries three different "color"-charged quarks, which combine into a neutral overall charge.

Weird quantum effects rule this level of matter. Quarks and mirror-image antiquarks pop instantaneously into being from a bubbling vacuum energy foam. Particles transmute into one another or tunnel through otherwise impenetrable barriers.

In these freaky places electricity may pass without resistance and super-fluid liquids can lose their viscosity and climb uphill.

The Proton
Two up quarks and one down
$2/3 + 2/3 - 1/3 = +1$

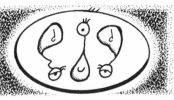

The Neutron
One up quark and two down
$2/3 - 1/3 - 1/3 = 0$

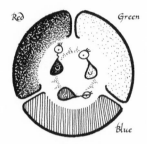

Red

Green

Blue

The red, green, and blue qualities of the
nuclear strong force combined, forming
a neutral white charge across a nucleon.

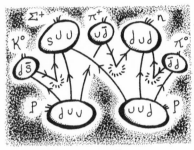

A Feynman diagram of two protons (p)
transforming into a neutron (n), a strange
sigma (Σ^+) particle, and a host of mesons
from spontaneous quark/antiquark creation.

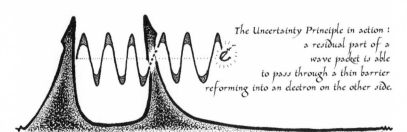

The Uncertainty Principle in action:
a residual part of a
wave packet is able
to pass through a thin barrier
reforming into an electron on the other side.

THE FOUR FORCES
holding the universe together

All things interact with each other through four universal forces, carried by four types of wave particles called *gauge bosons*.

The *electromagnetic* force is carried by *photons*. Light, x-rays, microwaves, and radio waves are all wrigglings of electromagnetic fields at different frequencies. This dual force attracts electrons to protons and causes most annihilations between matter and antimatter (*see below*). It is veritably the prime mover of all chemical reactions.

Acting over distances the size of a nucleus, the strong force binds quarks by exchanging eight types of *gluon*, the carriers of color charge. The strong force only affects quarks and gluons.

Particle decay is governed by the weak force, which acts over extremely short ranges. Changing a down quark into an up, for example, would transform a neutron into a proton and hence one element into the next. Carried by *W* and *Z vector bosons*, the weak force also allows neutrinos, very light fundamental particles, to perform their rare interactions with everyday matter.

Gravity, although the weakest force, works over almost infinite distances, extending its grip over all matter. Finding a quantum explanation for gravity has had its ups and downs, though a carrier boson, the *graviton*, has been tentatively described.

Positron (Antielectron) $E = Mc^2$ *Photons*

Electron

The electromagnetic force can operate over large distances : It shapes electron orbits and hence controls chemical behavior in atoms.

The strong force acts over distances the size of a nucleus, shaping and holding the quarks within together.

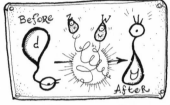

The weak force has a very short range indeed and is responsible for quark transformation and neutrino interactions.

Gravity acts across very great distances and connects all types of matter, from galaxies to the atoms of a cat.

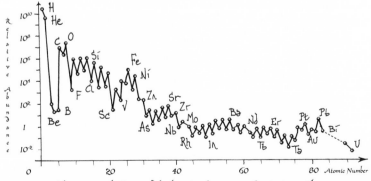

Relative cosmic abundances of the elements : The zig-zag is due to atoms with even numbers of protons and electrons being slightly more stable than ones with odd numbers.

QUARKS, LEPTONS, AND MESONS
fundamental families

The everyday matter of the universe is made of up and down quarks and their two *lepton* cousins, the electron and the neutrino. This is the first generation of indivisible fundamental particles. At higher energies, such as in cosmic rays, a second, similar but heavier, family of four is found, with *strange* and *charm* quarks and two more leptons, the *muon* and the *muon-neutrino*. At greater energies still a third, even more massive, family appears (*see opposite, top*). Each family also exists as its antimatter twin.

Composites of quarks are called *hadrons*, and take three forms: *baryons* made from three quarks, *anti-baryons* formed of three antiquarks, and finally *mesons*, which are quark/antiquark pairs. Hadrons alone are subject to the strong color force.

All particles have *spin*, a complex gyrotational symmetry. The quark and lepton matter families have spin $1/2$, magically having to turn around twice before looking the same. Particles with half spins ($1/2$, $3/2$, $5/2$ . . .) are called *fermions* (*see slugs below*).

All other particles have integer spins (0, 1, 2...) and are known as *bosons*. Mesons and the force-carrying gauge bosons are the primary examples. A selected few are on page 54.

The SPIN 2 mirror slug
has a half-turn symmetry

The SPIN 1 tumbling snail
rotates a full circle

..............BOSONS..............

The SPIN 1/2 moebius slug
twists through 720 degrees
FERMIONS

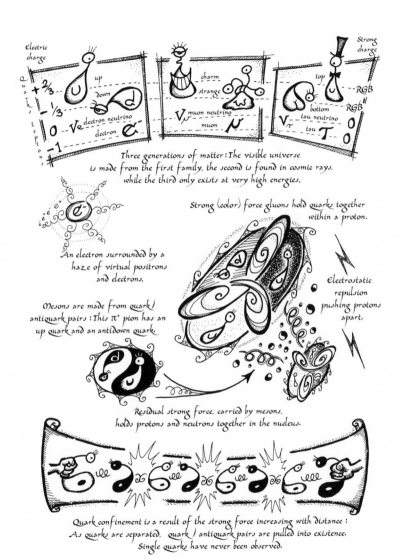

Electric charge

+ 2/3 up

− 1/3 down

0 V_e electron neutrino

−1 electron e

charm

strange

V_μ muon neutrino

muon μ

top

bottom

V_τ tau neutrino

tau T

Strong charge

RGB

RGB

0

0

Three generations of matter: The visible universe
is made from the first family, the second is found in cosmic rays,
while the third only exists at very high energies.

An electron surrounded by a
haze of virtual positrons
and electrons.

Strong (color) force gluons hold quarks together
within a proton.

Electrostatic
repulsion
pushing protons
apart.

Mesons are made from quark/
antiquark pairs: This π^+ pion has an
up quark and an antidown quark.

Residual strong force, carried by mesons,
holds protons and neutrons together in the nucleus.

Quark confinement is a result of the strong force increasing with distance:
As quarks are separated, quark/antiquark pairs are pulled into existence.
Single quarks have never been observed.

EXOTIC PARTICLES
and subatomic siblings

Cosmic rays hitting the upper atmosphere ionize atoms, creating cascades of subatomic particles. Visible as the polar auroras, branching streamers of baryons and mesons shed electrons, positrons, and γ-rays to cause further chain reactions. Short-lived particles transmute into stable, lower-energy ones.

Hidden structures are being revealed in experiments colliding protons and other matter at near light speeds in accelerators and cyclotrons where bubbles trace the curling paths of hundreds of exotic particles, colliding and transmuting. Some are higher-energy versions, or *resonances*, of others, once again blurring the distinctions between particles and waves.

Symmetrical patterns link together aspects of this orchestra. Called the *eightfold way* (after a Buddhist idea), whole families of relationships have been charted in octets (*opposite, lower left*) that correlate charges, spins, and other characteristics.

These hyperdimensional maps draw a new kind of periodic table as nature's strings reveal their subtle harmonics. So where does it all come from and who plucked the first note?

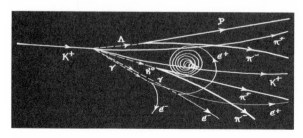

A high-energy cosmic ray sets off a cascade of subatomic particles.

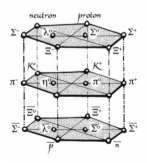

Eightfold Way Octets

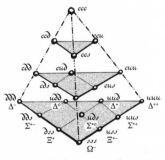

Spin 3/2 Baryon Family Tree

THE BIG BANG
from small beginnings

Once, according to a popular story, seemingly from nowhere, a pinpoint of intensely hot, immensely dense energy-matter appeared. This tiny *singularity* seeded our entire universe. Conditions were so bizarre we have only the vaguest idea of the exotic physics at the start of time.

With the young universe cooling fast and expanding at close to lightspeed, first photons, then quarks and leptons condensed out of the fizzing quantum vacuum like mist on a cold window.

After one millionth of a second the quarks had formed into hadrons, primarily protons and neutrons, as vast amounts of matter and antimatter wiped each other out. Soon, just a billionth part of matter was left, plus loads of gamma rays.

One second after the universe's birth, the temperature had fallen enough to crystallize whizzing neutrinos from the photons. Nucleosynthesis started about this time, with protons and neutrons fusing into helium, deuterium, and lithium.

Ten minutes later matter consisted of three parts hydrogen to one part helium. The universe was expanding incredibly fast, and after two hours there was no longer the density of neutrons to allow any heavier nuclei to form.

After 100,000 years the universe was cool enough to allow the electrons to settle into orbits around atomic nuclei.

Over the next few billion years matter drifted in dark clouds. In some places gravity gradually squeezed clouds into clumps, heat rising until nuclear fusion burst out, lighting the first stars.

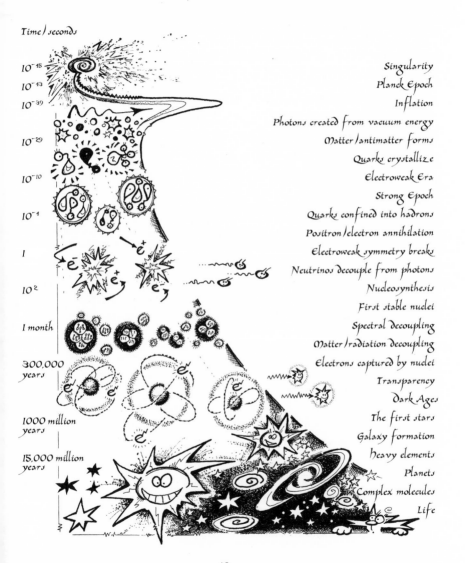

Time / seconds

10^{-45} — Singularity
10^{-43} — Planck Epoch
10^{-39} — Inflation

Photons created from vacuum energy
10^{-29} — Matter / antimatter forms

Quarks crystallize
10^{-10} — Electroweak Era

Strong Epoch
10^{-4} — Quarks confined into hadrons

Positron / electron annihilation
1 — Electroweak symmetry breaks

Neutrinos decouple from photons
10^{2} — Nucleosynthesis

First stable nuclei
1 month — Spectral decoupling

Matter / radiation decoupling
300,000 years — Electrons captured by nuclei

Transparency

Dark Ages
1000 million years — The first stars

Galaxy formation
15,000 million years — heavy elements

Planets

Complex molecules

Life

STELLAR FUSION
making the sun shine

We are born of stars. All of our atoms, except for primal hydrogen and helium, come from the massive furnaces of space.

Our sun is a second generation white dwarf. A huge nuclear fusion reactor, the enormous temperatures and pressures created by gravity squeeze four atoms of hydrogen together to form one of helium, releasing energy that reaches us as sunlight.

The hydrogen in stars eventually runs out and they collapse. Temperatures soar and the helium fuses into new elements, carbon, oxygen, and nitrogen. Stars the size of the sun fizzle out when the helium finishes.

For bigger stars it's a different story. When they reach this stage their greater mass ignites carbon fusion, scrunching into neon, oxygen, and magnesium. As the carbon is exhausted the core contracts again, this time producing silicon and sulphur. Intermediate elements are meanwhile painstakingly assembled by *slow neutron capture*, like a jigsaw that takes thousands of years.

With a further collapse the core is converted into iron, cobalt, and nickel. Since iron fusion uses up more energy than it gives out, the star literally expires and undergoes a drastic final implosion that ends in a cataclysmic supernova explosion.

In this immense cauldron many new elements are cooked up. Some are created by *rapid neutron capture* with bits and pieces being slammed together to form heavy atoms like uranium, while others are formed by *spallation*, where larger atoms are chipped into smaller pieces by ultra-high-speed debris.

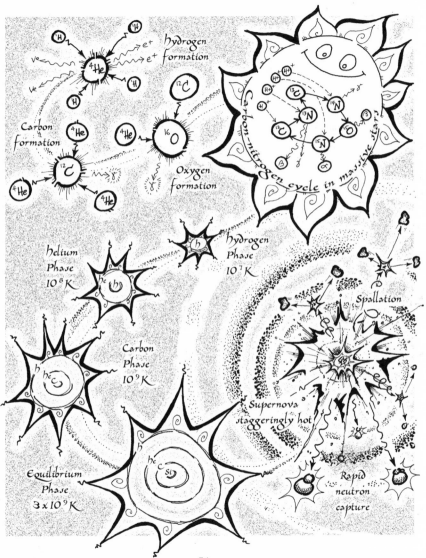

hydrogen formation

Carbon formation

Oxygen formation

Carbon-nitrogen cycle in massive stars

helium Phase 10^8 K

hydrogen Phase 10^7 K

Carbon Phase 10^9 K

Spallation

Supernova staggeringly hot

Equilibrium Phase 3×10^9 K

Rapid neutron capture

51

STRINGS AND THINGS
bubbles and branes

The more we probe into the wispy knottings of wave-matter stuff, the more there is to find. Currently beyond our experimental abilities, the colossal energies bound up in the quantum realm present a field day for adventurous mathematicians.

Attempts to find a Theory of Everything combining quantum mechanics, Einstein's relativity, and other universal patterns have produced many intriguing thoughts. Most invoke additional dimensions to the usual four of space-time.

Several types of *superstrings*, one dimensional threads that loop and vibrate across ten or so dimensions, have been mapped. Some quantum models that include gravity require eleven dimensions and *supersymmetries* to relate the elementary particles, connecting forces to phenomena. Others suggest that we are just one universe among an infinitude.

After a great deal of head scratching, the latest fad is that the knitting of superstrings and the pull of supergravity may be different viewpoints over a larger vista, the marvelous, mysterious M-theory. This pictures the universe as ripples in an infinitely vast, infinitely thin *p-brane* (like a membrane that spans many dimensions) stretching through hyperspace.

One idea is that the big bang was caused by branes colliding, our small universe being merely the interference patterns where they cross. Formed from hidden harmonies, perhaps in the end we are only as real as the holographic sparkles in a sunny pool, or the fractal rainbow swirls on the surface of a soap bubble.

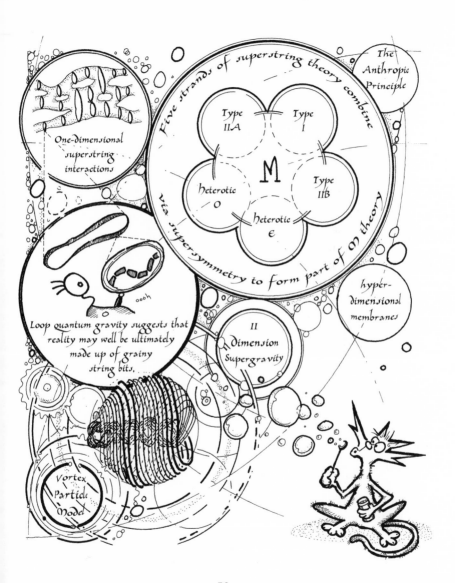

Five strands of superstring theory combine

via supersymmetry to form part of M theory

The Anthropic Principle

Type IIA

Type I

heterotic o

M

Type IIB

heterotic ε

One-dimensional superstring interactions

Loop quantum gravity suggests that reality may well be ultimately made up of grainy string bits.

oooh

11 dimension Supergravity

hyper-dimensional membranes

Vortex Particle Model

53

Constants and Hadrons

Electron mass	m_e	$9.1091534 \times 10^{-31}$ kg	
		$= 0.5110$ MeV	
Electron charge	e	1.602189×10^{-19} C	
		$= 4.8030 \times 10^{-10}$ esu	
Protomass	m_p	1.672648×10^{-27} kg	
		$= 1836.1 \times$ electron mass	
Neutron mass	m_n	$1.6749545 \times 10^{-27}$ kg	
Atomic mass unit	u	1.66054×10^{-27} kg	
Avogadro's no.	N_A	6.022045×1023 mol^{-1}	
Bohr radius	a_o	$0.52917706 \times 10^{-10}$ mol^{-1}	
Boltzmann const.	k	1.380662×10^{-23} J K^{-1}	
Faraday constant	F	9.648456×10^4 C mol^{-1}	
Gas constant	R	8.31441 J K^{-1} mol^{-1}	
H$_2$0 triple point	T_{tpw}	273.16 K	
Ice point temp.	T_{ice}	273.1500 K	
Mol. vol. gas	V_M^o	2.241383×10^{-2} m^3 mol^{-1}	
Perm. of vacuum	μ_o	$4\pi \times 10^{-7}$ H m^{-1}	
Permittivity const.	ε_o	8.8542×10^{-12} F m^{-1}	
Planck constant	h	6.626176×10^{23} mol^{-1}	
Planck length	l_p	1.616×10^{-35} m	
Planck time	t_p	5.319×10^{-44} s	
Planck mass	m_p	2.1777×10^{-8} kg	
Rcp. fine str. const.	$1/\alpha$	137.036	
Rydberg const.	R_H	1.097373×10^7 m^{-1}	
Speed of light	c	2.99792458×10^8 m s^{-1}	
1 angström (Å)		10×10^{-10} m	
1 atmosphere		101325 N m^{-2} (Pa)	
1 calorie (cal)		4.184 joules (J)	
1 celsius (°C)		273.150 Kelvin (K)	
°celsius		$5/9$ (°fahrenheit − 32)	
1 curie (Ci)		3.7×10^{10} s^{-1}	
1 erg		2.390×10^{-11} kcal	
1 esu		3.3356×10^{-10} C	
1 eV		1.60218×10^{-19} J	
1 eV / molecule		96.485 kJ mol^{-1}	
1 kcal mol^{-1}		349.76 cm^{-1}, 0.0433 eV	
1 kJ mol^{-1}		83.54 cm^{-1}	
1 wave no. (cm^{-1})		2.8591×10^{-3} kcal mol^{-1}	

Baryons	Symbol	Mass	Quarks	Charge	Spin
Proton	N^+	938	uud	+1	1/2
Neutron	N^0	940	udu	0	1/2
Sigma$^+$	Σ^+	1198	uus	+1	1/2
Sigma^0	Σ^0	1192	dus	0	1/2
Sigma-	S-	1197	dds	−1	1/2
Lambda^0	Λ^0	1116	dus	0	1/2
Xi^0	Ξ^0	1315	uss	0	1/2
Xi$^-$	Ξ^-	1321	dss	−1	1/2
Sigma$^+$	Σ^+	938	uus	+1	1/2
Delta^{++}	Δ^{++}	1231	uuu	+2	3/2
Delta$^+$	Δ^+	1232	duu	+1	3/2
Delta^0	Δ^0	1234	ddu	0	3/2
Delta$^-$	Δ^-	1235	ddd	−1	3/2
Sigma^{*+}	Σ^{*+}	1189	uus	+1	3/2
Sigma^{*0}	Σ^{*0}	1193	dus	0	3/2
Sigma^{*-}	Σ^{*-}	1197	dds	−1	3/2
Xi^{*0}	Ξ^{*0}	1315	uss	0	3/2
Xi^{*-}	Ξ^{*-}	1321	dss	−1	3/2
Omega$^-$	Ω^-	1672	sss	−1	3/2

Mesons					
Pi$^+$	π^+	140	$u\bar{d}$	+1	0
Pi^0	π^0	135	$u\bar{u}, d\bar{d}$	0	0
Pi$^-$	π^+	140	$d\bar{u}$	−1	0
Eta	η^+	547	$u\bar{u}, d\bar{d}, s\bar{s}$	0	0
Eta prime	η'	958	$u\bar{u}, d\bar{d}, s\bar{s}$	0	0
Kaon$^+$	K^+	494	$u\bar{s}$	+1	0
Kaon^0	K^0	498	$d\bar{s}$	0	0
Rho$^+$	ρ^+	770	$u\bar{d}$	+1	1
Rho^0	ρ^0	770	$u\bar{u}, d\bar{d}$	0	1
Omega	ω	782	$u\bar{u}, d\bar{d}$	0	1
Phi	ϕ	1020	$s\bar{s}$	0	1
K^{*+}	K^{*+}	892	$u\bar{s}$	+1	1
K^{*0}	K^{*0}	892	$d\bar{s}$	0	1
J/ψ	ψ	3097	$c\bar{c}$	0	1
Upsilon	Y	9460	$b\bar{b}$	0	1

Note: Over 200 baryons and 36 mesons are known.

CARBON CHEMISTRY

Name	Functional Group	Suffix	First Member Formula	First Member Name	General Formula
Alkane	$R-CH_3$	-ane	$H-\overset{H}{\underset{H}{C}}-H$	methane	C_nH_{2n+2}
Alkene	$\overset{H}{\underset{R}{}}C=C\overset{H}{\underset{R}{}}$	-ene	$\overset{H}{\underset{H}{}}C=C\overset{H}{\underset{H}{}}$	ethene	C_nH_{2n}
Alkyne	$R-C\equiv C-R'$	-yne	$H-C\equiv C-H$	ethyne	C_nH_{2n-2}
Alcohol	$R-\overset{OH}{\underset{H}{C}}-R'$	-anol	$H-\overset{OH}{\underset{H}{C}}-H$	methanol	$C_nH_{2n+1}OH$
Aldehyde	$R-\overset{O}{C}-H$	-anal	$H-\overset{O}{C}-H$	methanal	$C_nH_{2n-1}OH$
Ketone	$R\sim\underset{O}{C}\sim CH_3$	-anone	$H\sim\underset{O}{C}\sim CH_3$	methanone	$C_nH_{2n}O$
Carboxylic acid	$R-\underset{O}{C}-OH$	-anoic acid	$H-\underset{O}{C}-OH$	methanoic (formic) acid	
Amine	H_2N-R	-anamine	$CH_3-N\overset{H}{\underset{H}{}}$	methylamine	
Ester	$R\sim\overset{O}{C}\sim OR'$	-ate	$CH_3-C\overset{O}{\underset{OCH_3}{}}$	carboxylic acid + alcohol	
Amide	$R\sim\overset{O}{C}\sim NHR'$	-ide	$CH_3\overset{O}{C}\sim NHR'$	carboxylic acid + amine	
Ether	$R-O-R'$	-oxy... -ane	CH_3-O-CH_3	methoxy methane	
Cyclopropane	$H-\overset{H}{\underset{H}{C}}\diagup\overset{H}{\underset{H}{C}}\diagdown\overset{H}{\underset{H}{C}}-H$	Benzene	$H-C\diagup\underset{}{\bigcirc}\diagdown C-H$ or $\bighexagon$ or $\bighexagon$		

Branched alkanes :
branch name changes from -ane to -yl.

$$H_3C-\overset{}{\underset{CH_2}{}}-\overset{CH_2}{\underset{}{}}-\overset{CH}{\underset{CH_2}{}}-CH_3$$

3-ethyl hexane

$$H_3C-\overset{CH_2}{}-\overset{CH_2}{}-\overset{CH}{\underset{CH_3}{}}$$

ethyl branch: H_2C-CH_3

methyl branch: CH_3

2-methyl pentane

Prefix (no. of carbon atoms): Meth-1, Eth-2, Prop-3, But-4, Pent-5, Hex-6, Hept-7, Oct-8, Non-9, Dec-10

THE PERIODIC TABLE

Group	IA	IIA		IIIB	IVB	VB	VIB	VIIB	VIII
	1	2		3	4	5	6	7	8

Period

Period									
1	H 0 / **1** / **H** / Hydrogen / 1.00079 1310								
2	B 3 / **4** / **Li** / Lithium / 6.941 519	H 4 / **5** / **Be** / Beryllium / 9.01218 900							
3	B 11 / **12** / **Na** / Sodium / 22.9878 494	H 12 / **12** / **Mg** / Magnesium / 24.3050 736							
4	B 19 / **20** / **K** / Potassium / 39.0983 418	C 20 / **20** / **Ca** / Calcium / 40.0785 590		H 21 / **24** / **Sc** / Scandium / 44.9559 632	H 22 / **26** / **Ti** / Titanium / 47.867 661	B 23 / **28** / **V** / Vanadium / 50.9415 648	B 24 / **28** / **Cr** / Chromium / 51.9961 653	C 25 / **30** / **Mn** / Manganese / 54.9380 716	B 26 / **26** / **Fe** / Iron / 55.845
5	B 37 / **48** / **Rb** / Rubidium / 85.4678 402	F 38 / **50** / **Sr** / Strontium / 87.62 548		H 39 / **50** / **Y** / Yttrium / 88.9059 636	H 40 / **50** / **Zr** / Zirconium / 91.224 669	B 41 / **52** / **Nb** / Niobium / 92.9064 653	B 42 / **56** / **Mo** / Molybdenum / 95.94 694	H 43 / **55** / **Tc** / Technetium / 97.9072 699	H 44 / **Ru** / Ruthenium / 101.07
6	B 55 / **78** / **Cs** / Caesium / 132.905 376	B 56 / **82** / **Ba** / Barium / 137.327 502	57 - 70 Lanthanide series ✱	H 71 / **104** / **Lu** / Lutetium / 174.967 481	H 72 / **108** / **Hf** / Hafnium / 178.49 531	B 73 / **108** / **Ta** / Tantalum / 180.948 760	B 74 / **110** / **W** / Tungsten / 183.84 770	H 75 / **112** / **Re** / Rhenium / 186.207 762	H 76 / **Os** / Osmium / 190.23
7	H 87 / **136** / **Fr** / Francium / 223.02 381	H 88 / **138** / **Ra** / Radium / 226.025 510	89 - 102 Actinide series ✱✱	? 103 / **157** / **Lr** / Lawrencium / 262.110 ?	? 104 / **153** / **Rf** / Rutherfordium / 263.113 ?	? 105 / **157** / **Db** / Dubnium / 262.114 ?	? 106 / **157** / **Sg** / Seaborgium / 266.122 ?	? 107 / **157** / **Bh** / Bohrium / 264.125 ?	? 108 / **Hs** / Hassium / 269.134

✱

H 57 82 / **La** / Lanthanum / 138.906 540	C 58 82 / **Ce** / Cerium / 140.116 665	H 59 82 / **Pr** / Praseodymium / 140.908 556	H 60 82 / **Nd** / Neodymium / 144.24 607	H 61 84 / **Pm** / Promethium / 144.913 556	R 62 90 / **Sm** / Samarium / 150.36 540	B 63 90 / **Eu** / Europium / 151.964 548	H 64 94 / **Gd** / Gadolinium / 157.25 594	H 65 / **Tb** / Terbium / 158.925

✱✱

C 89 138 / **Ac** / Actinium / 227.028 669	C 90 142 / **Th** / Thorium / 232.038 674	T 91 140 / **Pa** / Proctactinium / 231.0356 ?	O 92 146 / **U** / Uranium / 238.029 385	O 93 144 / **Np** / Neptunium / 237.048 ?	M 94 150 / **Pu** / Plutonium / 244.064 ?	H 95 148 / **Am** / Americium / 243.061 ?	H 96 151 / **Cm** / Curium / 247.070 ?	H 97 / **Bk** / Berkelium / 247.070

IUPAC interim naming system for new elements : o-nil-(n), 1-un-(u), 2-bi-(b), 3-tri-(t), 4-quad-(q), 5-pent-(p), 6-hex-(h), 7-sept-(s), 8-oct-(o), 9-e